YOUR KNOWLEDGE HAS VALUE

Yvette Mhlanga

Epilepsy. Short Overview

GRIN Publishing

Bibliographic information published by the German National Library:

The German National Library lists this publication in the National Bibliography; detailed bibliographic data are available on the Internet at http://dnb.dnb.de .

Imprint:

Copyright © 2014 GRIN Verlag GmbH
Print and binding: Books on Demand GmbH, Norderstedt Germany
ISBN: 978-3-656-90411-3

This book at GRIN:

http://www.grin.com/en/e-book/293135/epilepsy-short-overview

GRIN - Your knowledge has value

Since its foundation in 1998, GRIN has specialized in publishing academic texts by students, college teachers and other academics as e-book and printed book. The website www.grin.com is an ideal platform for presenting term papers, final papers, scientific essays, dissertations and specialist books.

Visit us on the internet:

http://www.grin.com/

http://www.facebook.com/grincom

http://www.twitter.com/grin_com

According to census estimates, one in 103 of the UK population have epilepsy and it is usually misdiagnosed in 20–31% of cases (Stokes et al. 2004). Sisodiya & Sander (2004) suggests that with an optimal treatment routine 70% of the population could become seizure free. However, the reality is less encouraging since Moran et al (2004) found that only 52% of people with epilepsy were living without seizures. This reflection considers my experience in caring for a service user with learning disability (LD) and suffers from epilepsy. Furthermore, it will focus on the bio-medical aspects of epilepsy and how it has affected my service user and the treatment available. In addition, it will explore the meaning of quality of life and the psychological factors experienced by the service user in their daily life. Names of individuals have been changed to adhere to confidentiality, NMC (2008).

Causes of epilepsy are well documented, for example birth trauma, head trauma/injury, brain infections, and inherited brain disease. However, causes vary between individuals. People with epilepsy are believed to have a low seizure threshold therefore this makes them prone to seizures. This threshold is said to be influenced by any factors that triggers the seizures (Gomersall 2009). Gomersall (2009) suggests that epilepsy can be triggered by a number of different factors for instance, menstruation, tiredness, stress, too much alcohol boredom, photosensitivity and low blood sugar. Epilepsy affects different lobes of the brain therefore the lobe affected will influence the type of seizure experienced by individuals. Lear-Kaul et al, (2005) studied autopsy reports to identify the association between scientific variables and sudden unpredicted death in epilepsy while Walczak et al, (2001) had an observational study on 16,463 patients in three epilepsy centres over an average period of 4 years; both studies found that mortality rates for people with epilepsy are two to three times higher than those in the general population. In comparison these figures show that the prevalence of the condition of epilepsy in the UK is lower compared to the USA. This is because the UK national guidelines recommend that all individuals with epilepsy should have free access to an epilepsy specialist nurse, whereas in the USA individuals have to pay for their health cost.

As part of my placement I cared for Mary who was forty years-old with Learning disabilities. She was diagnosed with epilepsy when she was 15 years old, Mary experienced a lot of seizures and was on more than one anti-epileptic drug. After visiting the hospital with her for a review of her medication, the doctor explained the patterns of her seizure which tended to occur during her menstrual cycle. This explanation is verified by Kim et al (2010) who investigated seizure frequency in 79 premenopausal women with epilepsy aged 15-44 years and discovered that over 46% of them had a greater occurrence of seizures during their menstrual cycle. In reflection I felt empathy for Mary as she had to contend with these catamenial seizures every month when she has her menstrual cycle. This challenged me to learn more on how to care for her as I understood her condition more.

Foundation for People with Learning Disabilities (2008) stated that seizures and its triggers may sometimes go unnoticed in people with LD as it can be difficult to distinguish from other behaviour associated with Learning Disabilities. In Mary's case it was thought to be challenging behaviour rather than epilepsy, through this experience I have learnt that epilepsy and seizures can be challenging especially when it is with people with Learning disability.

According to Schmitz *et al* (2010) patients with epilepsy show a higher risk of depression compared to the general public. They also stated that women were mainly at risk compared to men due to sex hormones which are known to contribute to epilepsy and depression. This is supported by Epilepsy Research UK (2009) who state that depression is common in people with temporal lobe epilepsy (TLE) which affects 20-40% compared to 7-12% of the general public which according to their research is due to cytokine known as interlukinlbeta (ILIB) which increases among people with TLE.

Significantly, epilepsy is a public health concern: with severe physical and psychological consequences. Ba-Diop et al (2014) state that it is serious and can cause traumatic injury, premature death, and mental health illness such as stress or depression. In the UK public health emphasis is on individuals to maintain a healthy lifestyle in order to reduce health related conditions. Gates and Barr (2009) suggested that maintaining a healthy lifestyle in people with epilepsy is important for example exercise, diet, regular sleep pattern and controlled alcohol intake which will possibly reduce the stress. However, Daley et al (2009) suggested that only exercise reduce the symptoms of depression. Scientifically, depression is commonly treated with anti-depressants or psychotherapy. However, some people may prefer alternative approaches. Through caring for Mary I noticed that exercises helped her to cope with her depression. However, for patients with Mary's condition it can be challenging for them to maintain a routine of exercises as a seizure can occur at any time (Kim et al., 2010). Therefore, places close to water should be avoided.

Some individuals prefer using complementary therapies rather than anti- epileptic drugs. These therapies are mainly acupuncture, aromatherapy, biofeedback, herbal treatment and homeopathy. Although these therapies are believed to work in some individuals, Nadkarni & Saxena (2011) recommend that they be used to complement anti-epileptic drugs due to lack of substantive scientific evidence that they can work on their own.

Nonetheless, patients opt for alternative therapies because unlike medication they have no known side effects.

Due to Mary's condition she was using different services for treatments to manage her epilepsy and disability which included anti-epileptic drugs as well as complementary therapies. Mary also used a drug called rectal diazepam when she was experiencing a series of seizures without fully recovering. National Institute for Clinical Excellence (2008) suggested that rectal diazepam is the fastest to be absorbed in the system compared to other oral medication. However, the drug has its own side effects despite being effective. It is a short-term effect drug that causes drowsiness and unsteadiness. It is thought that it causes anxiety and hallucinations on individuals (Stefan & Theodore, 2012). This can affect the patient's mental health if left unchecked or if the patient had undetected mental health issues. A study conducted by Berman et al (2005) suggests that the side effects of diazepam caused suicidal and self-aggressive acts.

Another drug that Mary was prescribed was the Buccal Midazolam drug. According to Marshall (2007) the drug is as effective as the rectal diazepam. However, it is still unlicensed thus rectal diazepam remains the first option. Marshall (2007) states that although buccal midazolam is unlicensed doctors are allowed to prescribe the drug when they agree with the patient, family and carers or when they feel it is the best option for the patient. The research Marshall (2007) carried out shows that most families and carers prefer buccal midazolam compared to rectal diazepam. However, these findings could be biased as it does not mention the side effects of the drug and I suspect that its aim was to appraise the drug.

Through research I have learnt that epilepsy can be life threatening and seizures can lead to health complications. Therefore, it is vital for nurses to be in supply of drugs such as rectal diazepam and buccal midazolam when caring for patients with epilepsy.

One of my concerns while caring for Mary was her social interaction. Most individuals with epilepsy will face a range of concerns with their quality of life, for example social identity. Quality of life has been linked to a broad range of concerns. Jacoby, *et al* (2005) states that the stigma associated with having epilepsy is common to many cultures which normally results in individuals experiencing a negative effect on their social identity. Jacoby, *et al* (2005) believes that once a person's social identity is destroyed it is almost impossible to reacquire the status of normality. This results in isolation of which people like Mary will struggle with self-esteem as asserted by Jacoby, *et al* (2005). Practically, this will be evident through diminishing social life, going out, fewer friends, and impeded life chances.

Another important social aspect affected by epilepsy is driving. According to DVLA (2009) people with epilepsy need to be free from seizures for at least three years to be able to obtain a license. This is for public safety including the people with epilepsy since seizures do not usually give any warning. Furthermore, employment is another social and economic factor that people with epilepsy face, according to Epilepsy and Employment Scotland (2008). Numerous surveys have shown high rates of unemployment or underemployment in persons with epilepsy (Preedy & Watson, 2010).

Reflecting on practice most of the individuals I have supported who are diagnosed with epilepsy seems to be getting their seizure occurrence under control due to regular check-ups and medication reviews they are attending. While on placement I have also noticed that there is a lot of awareness in terms of epilepsy. For example, I attended a day session on epilepsy for me to be able to support service users who had epilepsy. I feel that it is good practice for everyone to have some understanding of epilepsy before supporting any patients or service users. Hayes (2004) assert that the key role of nurses towards epileptic patients is advisory and educational. She further states that facilitating and assisting adults to self-manage their condition are skills passed on from training sessions. However, there are numerous barriers, including professional's own knowledge limitations, time restraints and the patient's own engagement with the nurse.

Through the teaching session I had on placement I have learnt that although there is still a lot of stigma attached to epilepsy in the general public those in health and social care have a better understanding of epilepsy and are less likely to stigmatise it. I have also learnt that epilepsy and seizures can be difficult to notice in people with learning disabilities hence people supporting individuals with learning disabilities should be observant and have effective communication skills. Gates and Barr (2009) suggest that individuals, families and carers should be educated about epilepsy to get better knowledge and raise awareness. There is a lot of awareness on epilepsy through different websites, for example Epilepsy Action, Epilepsy Research and National Society for Epilepsy. However, this can only be limited to people who access the internet therefore more public health promotion through different ways should be applied, for example television advertisement, radios and local newspapers.

This assignment has looked at the bio-medical aspects of epilepsy and how it affects individuals, the treatment available to minimize seizures including the complementary therapies. The psychological aspects of epilepsy have been explored. The way it affects individuals resulting in changes of behaviour, frustration, stress and in some cases depression. This assignment has also outlined how depression is more common in people with temporal lobe epilepsy compared to the general public. The social aspects of epilepsy which are being unable to drive, employment issues and individuals not being able to fulfil their social roles within the community has been explored. This assignment has also explored on the social policy and public health and how it relates to health and social care.

References

Ba-Diop, A., Marin, B., Druet-Cabanac, M., Ngoungou, E., Newton, C., & Preux, P. (2014). Epidemiology, causes, and treatment of epilepsy in sub-Saharan Africa. *The Lancet Neurology, 13*(10), 1029-1044. doi:10.1016/s1474-4422(14)70114-0

Berman, M., Jones, G., & McCloskey, M. (2005). The effects of diazepam on human self-aggressivebehavior. *Psychopharmacology, 178*(1),100-106.doi:10.1007/s00213-004-1966-8

Daley, A., Jolly, K., & MacArthur, C. (2009). The effectiveness of exercise in the management of post-natal depression: systematic review and meta-analysis. *Family Practice, 26*(2), 154-162. doi:10.1093/fampra/cmn101

DVLA (2009) *for medical practitioners: at a glance guide to the current medical standards of fitness to drive.* DVLA. www.dvla.gov.uk

Foundation for People with Learning Disabilities (2008*) Mutual Caring: supporting Mutual Caring amongst Older Families that Include a Person with a Learning Disability. A Briefing Note for Policy Makers, Commissioners and Services. London: Mental Health Foundation*

Gates, B., & Barr, O. (2009). *Oxford handbook of learning & intellectual disability nursing.* Oxford: Oxford University Press.

Gomersall S (2009).*Providing training to improve the understanding of the epilepsies,* Epilepsy Awareness.

Hayes, C. (2004). Clinical skills: practical guide for managing adults with epilepsy. *British Journal of Nursing, 13*(7), 380-387. doi:10.12968/bjon.2004.13.7.12681

Jacoby, A., Snape, D., & Baker, G. (2005). Epilepsy and social identity: the stigma of a chronic neurological disorder. *The Lancet Neurology, 4*(3), 171-178. doi:10.1016/s1474-4422(05)70020-x

Kim, G., Lee, H., Park, H., Lee, S., Lee, S., & Kim, Y. et al. (2010). Seizure exacerbation and hormonal cycles in women with epilepsy. *Epilepsy Research, 90*(3), 214-220. doi:10.1016/j.eplepsyres.2010.05.003

Lear-Kaul, K., Coughlin, L., & Dobersen, M. (2005). Sudden Unexpected Death in Epilepsy. *The American Journal Of Forensic Medicine And Pathology, 26*(1), 11-17. doi:10.1097/01.paf.0000154453.58795.18

Marshall, T. (2007). A systematic review of the use of buccal midazolam in the emergency treatment of prolonged seizures in adults with learning disabilities. *British Journal of Learning Disability, 35*(2), 99-101. doi:10.1111/j.1468-3156.2007.00441.x

Moran, N., Poole, K., Bell, G., Solomon, J., Kendall, S., & McCarthy, M. et al. (2004). Epilepsy in the United Kingdom: seizure frequency and severity, anti-epileptic drug utilization and impact on life in 1652 people with epilepsy. *Seizure, 13*(6), 425-433. doi:10.1016/j.seizure.2003.10.002

Nadkarni, V., & Saxena, V. (2011). Nonpharmacological treatment of epilepsy. *Ann Indian Academia of Neurology, 14*(3), 148. doi:10.4103/0972-2327.85870

National Institute for Health and Clinical Excellence. (2008). *Nursing Standard, 22*(50), 30-30. doi:10.7748/ns2008.08.22.50.30.p4587

Nursing & Midwifery Council (2008) The code: standards of conduct, performance and ethics for nurses and midwives. Nursing& Midwifery Council. London

Nursing & Midwifery Council Code of Professional Conduct. (2008). *Nursing Ethics, 9*(6), 674-680. doi:10.1191/0969733002ne561xx

Preedy, V., & Watson, R. (2010). *Handbook of disease burdens and quality of life measures.* New York: Springer.

Schmitz, E., Robertson, M., & Trimble, M. (2010). Depression and schizophrenia in epilepsy: social and biological risk factors. *Epilepsy Research, 35*(1), 59-68. doi:10.1016/s0920-1211(98)00129-6

Shorvon, S. (2010). *Handbook of epilepsy treatment.* Chichester, West Sussex, UK: Wiley-Blackwell.

Sisodiya, S., & Sander, J. (2004). Epilepsy: management. *Medicine, 32*(10), 52-56. doi:10.1383/medc.32.10.52.51494

Stefan, H., & Theodore, W. (2012). *Epilepsy.* Edinburgh: Elsevier.

Walczak, T., Leppik, I., D'Amelio, M., Rarick, J., So, E., & Ahman, P. et al. (2001). Incidence and risk factors in sudden unexpected death in epilepsy: A prospective cohort study. *Neurology, 56*(4), 519-525. doi:10.1212/wnl.56.4.519